⟨NS⟩

The REDESIGN of TOMORROWLAND

by Bert

L.N. Smith Publishing
Madison, Wisconsin

Published by:

⟨NS⟩ L.N. Smith Publishing
3818 Birch Avenue
Madison, WI 53711

Copyright © 2009 by L.N. Smith

Dedicated to
Walt Disney and Uncle Albert

Contents

Prologue

Pinocchio would have been wise to listen to Jiminy Cricket; the people of the 1700s would have been wise to listen to Benjamin Franklin; and my adolescent children would be wise to listen to their parents. But heeding good advice is not a certainty, which is why adults must sometimes lead their children by the hand. Yet, in a world of nearly seven billion people, just *who* are the adults and *who* are the children? Age is not necessarily the qualifier.

I'm a 45-year-old man, and I make no claim to be wise. I'm a grown-up child who can play *Guitar Hero* better than his two teenage children and all of their friends. It's a useless skill, I know, but I can't deny my fondness for it. This computer game did me a great service: it functioned as my muse while I redesigned Disney World's Tomorrowland.

You see, there are still Benjamin Franklins in the world we should be listening to. They are the experts holding title to some of the groundbreaking discoveries that will soon reshape life on the planet. But the news of their discoveries has been slow to disseminate.

"How can this be?" you might ask. "If they've discovered the truth, then why aren't people pouncing on it?"

Well, the answer is, that people — even the experts — have blind spots. Time and time again, the truth has paraded before their eyes, though they've failed to see it. Just consider the following.

When Copernicus presented the math that refuted the concept of an earth-centered universe, few of the clerics and natural philosophers of his time accepted the idea. Even the esteemed Galileo couldn't sway this descendent panel of experts some seven decades later. Likewise, in the early 1900s, almost every physicist rejected Einstein's notion of bundled light.

When a breakthrough discovery threatens the status quo . . . when the glimmer of a new paradigm threatens to sweep the work of countless experts into the dustpan of time . . . the reaction tends to be *resistance*.

I suppose some scholars would take issue with me on this point, claiming they are sufficiently open-minded and intellectually detached to accept any new idea that has merit. But therein lies the problem. Merit is a judgment call that is cloistered within the prevailing paradigm. And a paradigm, by definition, means that judgment is confined to the cloister. Furthermore, these experts are entirely human. Their great credentials are predicated on institutions deeply rooted in the established paradigm. Only the most maverick of scholars can be early to adopt a revolutionary idea. Perhaps this is why the world's truly provocative insights have often arisen outside the halls of academia.

My purpose here is to inform readers that there are some revolutionary ideas now circulating. A few lucky truth-seekers have spotted them already . . . and I might just be one of them. This book is my attempt to share what I've learned.

Foremost among my discoveries is the newfound perspective that the human species is on a perilous path. It is nearing a juncture in its evolution where its two most probable routes are either straight up or straight down. The first and more desirable path offers a swift ascent into a new and glorious Golden Age. But the more likely path brings a catastrophic plunge into the darkest of Dark Ages. People need to get wise, and they need to do it in a hurry. Their unpleasant fates—and especially those of their children—will be sealed during the coming decades unless changes are made.

The Tomorrowland of these pages suggests a few such changes and is meant as fuel for debate. It stems from my imagination and has drawn free commentary from the members of my family, whose remarks I've sprinkled throughout. My hope is that their "expert" testimony won't deter you from undertaking further investigations on your own, but rather, I hope it will alert you to the rhetoric of the prevailing paradigms.

So, please, if you'll humor me. Sit back and enjoy my attempt at edutainment. From *retro*-future to *ready*-future, welcome to the new Tomorrowland.

CHAPTER 1

The "Be a Real Boy" Statue
(new feature)

The giant teacups of the hatter's Mad Tea Party spin behind
me, as I stride away to the whistlings of the *Unbirthday Song*.
I'm walking through the Magic Kingdom, from Fantasyland
into Tomorrowland.

A sculpture catches my eye under the shade of a tree in the
center of the walkway. This north entrance to Tomorrowland
has a new icon, and I stop to take a look at it.

The bronze sculpture is a statue of Benjamin Franklin
carrying a lantern and holding Pinocchio by the hand. The
inscription on its plaque is a rework of a proverb made famous
by Franklin.

> Early to learn, and see truth through the lies, makes
> a person . . . a nation . . . a world . . . healthy, wealthy,
> and wise.
>
> —L.N. Smith

CHAPTER 2

Charming's Grand Prix Raceway
(formerly *Tomorrowland Indy Speedway*)

<u>TYPE</u>: outdoor drive-it-yourself miniature car ride
<u>DURATION</u>: 3½ to 5½ minutes
<u>CAPACITY</u>: 1 to 2 riders per car
<u>HEIGHT REQUIREMENT</u>: at least one rider must be 52"
<u>LOADING SPEED</u>: moderate
<u>AVERAGE WAIT</u>: 3½ minutes per 100 people ahead in line
<u>RECOMMENDED AGES</u>: 3 and up
<u>DESCRIPTION</u>: Miniature electric cars travel around a track
at four to seven miles per hour. Sensors and positioning
devices prevent vehicles from colliding, and drivers can
switch lanes if there's room. Cars are self-piloting when
the controls are not being operated.

I enter the pavilion and join the line, where a succession of
illustrated signs tells a story to quicken the wait.

> Once upon a time in a faraway land, there was a
> tiny kingdom—peaceful, prosperous, and rich in
> romance and tradition. In the royal palace, there
> lived the widowed king and his young son, Adrian.
> Although the king was a devoted father who
> helped to raise his son, still he thought the boy
> needed a woman's care. And so he searched far and
> wide for a governess.

To the prince's good fortune, the candidate his father selected was a woman of superior intelligence and excellent character. She instilled in the boy her compassion, her confidence, her work ethic, her humility, her sense of humor, and her mastery of reading, writing, and arithmetic. Adrian couldn't help but grow to become a fine young man.

On his seventeenth birthday, the governess advised him to travel the world and see for himself the various ways in which people are governed. But when his father overheard the suggestion, he flew into a rage. "The boy doesn't need another mentor," he shouted, and he fired the governess on the spot.

Two weeks later, Adrian and a personal attendant waited until after dark and snuck from the palace. They loaded their provisions onto horses and left the kingdom, riding for many days until they eventually reached a sprawling metropolis.

This new city was much bigger than Adrian's home kingdom. Here, there were salons filled with men and women of elegance and pedigree, who, like himself, knew how to waltz and engage in witty conversations. For the first time in his life, Adrian was in the company of people he could relate to, and the experience was exhilarating. But even more intoxicating to Adrian were the women in their fine dresses and heavenly fragrances. It wasn't long before he was head-over-heels in love with the most beautiful and flirtatious belle of them all. And she seemed to return his affections . . . that is, until the day of the annual harness race.

On the city's biggest day of the year, the wagerers favored Adrian to cross the finish line first, and his horse and sulky held a commanding lead going into the final turn. But that's when fate intervened. A small boy fell from the grandstand onto the track, prompting Adrian to veer his vehicle to the side and shield the youngster from harm. The other drivers zoomed past him at top speed, and Adrian finished in last place. Many of the spectators booed Adrian, tearing up their tickets and grousing, "All for the life of a peasant child."

There were rumors of foul play, and Adrian was accused of forfeiting the race on purpose. Yet the biggest blow to Adrian that day was the loss of his first love, Madeleine. She became newly infatuated with the first-place finisher of the derby.

Adrian left the city with a heavy heart and a tarnished name. He vowed to never again be fooled by a woman whose advances were disingenuous, and he devised a clever strategy for mending his reputation. His first act was to send home his personal attendant. Then, he traveled alone into the mountains.

Upon reaching a high-elevation kingdom, Adrian introduced himself as the eldest son of a wealthy merchant from a faraway land. And when the nobility eventually pressed him for the details of his fortune, he explained that he had been disinherited because of his refusal to marry a woman he didn't truly love.

Adrian's lie worked better than he could have imagined. Not only did he brand himself as an unfit suitor to the ladies (making him no competition to the gentlemen) but he firmly established himself as a person of high principles. Adrian became a trusted confidant to everyone in the palace, including the king, who took him under his wing and trained him in the ways of administration.

The women of the courts developed a secret name for Adrian, referring to him as *The Charming Pauper* (which they quickly shortened to *Charming*).

Charming became the kingdom's champion downhill skier, and he was soon known throughout the realm by his nickname.

But then, on a moonlit winter's evening during a celebration of one of his ski victories, the enraptured queen cornered Adrian on a palace balcony and tried to kiss him. Adrian dodged her advances and escaped to his room, where he immediately packed his belongings and fled the kingdom.

Charming followed a mountain stream to the sea, where he entered a bustling port city.

The merchants of this seafaring kingdom were well traveled, and many of them had been across the ocean to the United States — a young nation with an experimental form of government called a republic. These merchants wanted their home kingdom to be a republic too, so they petitioned their king to relinquish his powers and yield to a document similar to the U.S. Constitution.

Charming supported the idea of reform, but he also supported the king, who insisted that self-governance, though a worthy aim, should not be thrust upon the people so abruptly. Such a radical change, he thought, should proceed in carefully measured steps to ensure a peaceful and orderly transition.

But the merchants wanted the new government right away, so they schemed an overthrow of the king.

They called a secret meeting and invited Charming to attend.

The conspirators planned to sink the king's boat during the annual sailing race, in which Charming had won the dubious honor of being the shipmate for the clumsy king. Charming's life would be spared if he pledged his loyalty to the coup d'état.

So Charming faked his loyalty and secretly hatched a plan to save the king. Charming would make the king's boat faster than all the others so that none of the co-conspirators could catch them.

Charming's first idea was to equip the king's sailboat with stabilizing pontoons that would double as power generators, harnessing the sea's natural wave energy to crank a propeller. But the system, according to his calculations, would add only a few extra knots to their speed, and it would fail completely if the sea was unusually calm on the day of the race. So the prince opted for his second idea. He hid a small boiler in the boat's hull and piped its steam and flue gases into a hollow cylinder mounted beneath the keel. This jet-type engine would motor them along in all kinds of weather, while operating silently below the waves.

Charming and the king took an early lead in the race and extended it throughout. They rang the Finishers' Bell a full two hours before any other racers, then kept on traveling along the coastline until they reached the next kingdom. There, the king took refuge and tried to govern from afar.

Meanwhile, Charming ventured on, determined to find a principality that offered him what he really wanted: access to an extensive library.

Word of Charming's exploits were, by now, well known across the land. So, when he entered a modest village set among farm fields and pastures, he was greeted like a hero. People rushed into the streets to welcome him, and they were quick to show him their new mechanical marvel — an invention that was sure to replace the horse and buggy: the automobile.

Charming was invited to live with the royal family in the palace. And there, he trained himself to be a driver for the kingdom's first ever Grand Prix Horseless Carriage Race.

Charming won that first race by a wide margin and captured the hearts of citizens, including the heart of the king's only child — the eight-year-old Princess Caroline. Caroline loved Charming, and she pleaded with her parents to let him serve as her private tutor. Charming jumped at this opportunity, for it gave him full use of the palace library.

During the cold winter months, the social gatherings were filled with talk of reform, as merchants and nobles lobbied their king for a representative form of government. The king listened and seemed open to the idea, so Charming entered the discussions and offered a stepwise plan for implementation. Everything was going well until the king finally comprehended that he would be replaced by an elected president, meaning his sovereign powers would be stripped away.

Immediately, the king's mood soured. He ordered all conversations on the topic to cease, and he grew distrusting of Charming and the other reformers.

The king eventually imprisoned all dissenters in the castle dungeon, except for Charming, whom he placed under house arrest and confined to a room in the palace tower. From there, Charming continued to perform his duties as Caroline's tutor, happily reading the books she smuggled to him from the library.

Each captive was to be held until publicly denouncing his support for the republic. And one by one, each subversive recanted, until Charming was the only one left under lock and key.

The stalemate between Charming and the king came to a head with the approaching date of the second Grand Prix race; the king felt considerable pressure from his subjects to release their former champion and let him defend his title.

The king made a daring decision. He offered Charming the hand of his daughter Caroline in marriage in exchange for Charming's pledge of loyalty to the crown. The king believed that Charming would make an excellent successor and that this young man's foolish notions of reform would fade once he took the throne.

Charming accepted the king's terms and agreed that the promised wedding to Caroline should be kept a secret until her sixteenth birthday. But Charming also insisted that his public declaration be made after the race rather than before, so that, if all went well, he could deliver his statement from the winner's circle. Charming was perpetrating another deception.

On the day of the race it was indeed Charming who crossed the finish line first, but he didn't stop his automobile there. Instead, he sped from the town without looking back.

Upon reaching the next territory, he sent a note to his father and asked for permission to return home. Charming was ready to settle down and help with the administration of his father's kingdom in preparation for his own ascension to the throne someday.

Charming had a plan for the incremental conversion of his kingdom to a republic. First, he would strengthen the justice system and encourage a free press. Then, he would introduce his population to the concept of casting ballots when deciding an issue, clearly posting all results and enacting the wishes of the majority. Once his people understood and trusted the process, he would then unveil for them a charter, similar to the U.S. Constitution, that would become the framework for their new government. Charming would gradually yield his authority to an elected congress, then give up his throne completely once a freely elected head of state was in place.

When the prince reached his father's palace, tired and dusty from the long ride in the motorcar, he learned of the ball to be held in his honor that evening. The prince was not pleased, and he was especially put off by his father's transparent plot to introduce him to every maiden in the kingdom. Charming did not believe that a woman existed anywhere in the world who could satisfy his unique tastes. Physically, he was attracted to the refined and perfumed women of the courts. Yet intellectually, these same women repulsed him with their varied vanities and ignorances. Charming prepared himself for a long evening of uninteresting girls. He never imagined the vision of beauty and grace that would appear before him on the terrace outside the ballroom.

Cinderella arrived late and alone. She was not accompanied by a chaperone or advocate to forward her interests, and she moved with an uncommon elegance. Her gown was the most extravagant the prince had ever seen, and she bore a manner that was both sincere and unassuming.

Charming ignored his royal etiquette and marched past the two awkward sisters who were then curtsying to him. He crossed the ballroom, with his eyes transfixed, then offered his hand to the new arrival.

Cinderella and Charming danced through the great hall and then outside into the gardens, where their eyes traded starlight and moonbeams as they twirled in each other's arms . . . that is, until the clock chimed midnight.

And you know the rest of the story. The prince and Cinderella were eventually married and lived happily ever after.

But what you might not know is that the two of them successfully converted their kingdom to a republic, diminishing their own roles to those of mere figureheads for a bygone era.

They established this Grand Prix Raceway — the ongoing road rally — to celebrate Charming's youthful adventures.

An awning covers the pits, where empty "race cars" move slowly, bumper-to-bumper, in a single-file line between stripes on the pavement.

I'm released into the area and board a car, where I listen

to the sound of an imaginary pit crew, as it supposedly pushes my unstarted roadster.

Soon, my vehicle and three others splay forward to a starting line.

"Ladies and gentlemen, start your engines!" calls an announcer.

My vehicle roars with a *vroom,* and I pump the accelerator to keep it revving. The hood trembles as if there's real torque beneath it.

Lights cascade down the starting pole, and the cars leap forward simultaneously. I press the gas pedal to the floor and quickly overtake some of the drivers ahead of me, cruising easily past them through this "countryside," which includes real sheep feeding on the grass beside the road.

I eventually take the checkered flag, and my vehicle silences itself immediately. It merges with the other cars into a single lane and then moves under an awning to an unloading area. The dashboard displays my time, comparing it to those of today's other drivers.

During my exit from the ride, there's one more storytelling sign.

> To this day, the people of Charming and Cinderella's republic remain grateful to the United States for blazing the trail of freedom and democracy. As a way of saying thank you, they want to share some of the wisdom they've acquired on their own.
>
> Please step up to a kiosk along the path and answer a few questions. You might just win a family pass to the VIP lounge in the skybox over the bleachers.

I stop beside a computer touchscreen and take its quiz. Screen One shows a picture of an overweight mouse named

Gus wearing a sweater that is bunched up around his face and neck.

> *How many health care systems exist in the United States?*
>   *(a) one*
>   *(b) two*
>   *(c) three*
>   *(d) four*

The correct answer is (d): *The United States has private insurance for the wealthy/employed, Medicare for seniors, VA insurance for veterans, and bankruptcy/Medicaid for the uninsured/poor. The total cost per person is by far the highest in the world, yet the overall health of the population is nothing to brag about.*

Screen Two features a mouse named Jaq waving his hat in the air.

> *When will the health care problem be solved?*
>   *(a) when "human health" becomes the essential measure, not profits for the food and health industries*
>   *(b) when research dollars cease to be directed by special interest groups, like agribusiness and pharmaceutical companies*
>   *(c) when health care providers heed the quality research that already exists*
>   *(d) all of the above*

The correct answer is (d): *The revolution in health care will begin when people learn the truth about nutrition and start applying it.*

Screen Three shows Jaq again, but this time he looks to be deep in thought.

Why is the world so polluted?
   (a) consumer apathy
   (b) weak governments
   (c) corporate greed
   (d) a simple accounting error

The correct answer is (d): *Balance sheets ignore "natural capital," assigning it a value of zero by default. This omission from the bookkeeping is a grave error: the environment and its ecosystems are among the most valuable assets on earth, certainly worthy of better than null treatment.*

I've achieved a perfect score on the quiz, though I'm denied a pass to the VIP lounge by the onscreen roulette wheel.

"The proper role of a free nation is to behave like a loving parent and help the less-free nations grow more mature . . . keeping in mind, of course, that the road to good parenting is seldom smooth, and that setting a good example is arguably the best teacher."   —Bert

CHAPTER 3

The "Conscience Be Your Guide" Statue
(new feature)

Outside the Grand Prix Raceway—at the opposite end of the tree line from the *Be a Real Boy* statue—stands another similar sculpture. This one features a woman carrying a lantern and holding the hand of a small ape. The plaque reads:

Dedicated to
Sue Savage-Rumbaugh
and her boy Kanzi

My attention is diverted from the statue, however, by a raspy voice from behind me. "Beware!" it says forebodingly.

An old woman, dressed in rags, swings her arms wildly, as if sweeping spirits from the sky.

"Beware," she shrieks again, and the people around me stop to observe her. "Be —" she clutches her throat and hacks out a cough. "Uh-hum, that's better." Her voice sounds normal now. "Listen up, people. Gather 'round; free yourself from the lies. The march of ides is upon you, so heed these words from the wise:

> Human rights and equal rights
> Are wholly different things.
> Kids have one but not the other.
> It's the same for modern kings.

Those of us who see it clear
Must work to lead the way,
'Cause no one thrives when playground rules
Are set by kids at play.

So human rights, yes, but equal rights, no;
Just listen to your sages.
The march of ides is upon you.
Your intellect develops in stages.

"And speaking of stages," she continues to chant. "Be quick
to the Tomorrowland Stage. For there you'll meet an ape
named Kanzi and discover the debate soon to rage."

The old woman turns and again swipes at the air, creating
an avenue for herself through the crowd. She hurries away
through it, crying, "Beware! Beware!"

I cross Tomorrowland and pass beneath the two attractions
that remain unaltered by my redesign — the Astro Orbiter
and the Tomorrowland Transit Authority. I then find a place
to stand where I can view the Tomorrowland Stage.

"There's nothing new here, Bert. The stages of intellect have
been studied for more than a century."

—Elsa, Ph.D. Psychology

"Yes, but the stages need to be common knowledge to more
than just a few professors, because every human relationship
is affected by them."   —Bert

CHAPTER 4

The Tomorrowland Stage
(new show)

The stage near Space Mountain is fronted by a giant video screen, which at the moment plays a commercial for a new television series. The series is called *Sculpt Me Beautiful,* and its creator is a wealthy investor who teamed up with the Disney Company and a local medical center to form an integrative health spa at one of Disney's resorts. Many of the spa's high-profile clients are either celebrities or former contestants of other weight loss shows, like *The Biggest Loser,* where past efforts have failed to yield a lasting body that is truly fit and trim. But the participants of this show always do better, and they do so without suffering the abuses of hunger, over-exercise, or needless surgeries. Their make-overs result in stunning transformations of both health and appearance, and these outcomes are flaunted through games and fitness challenges that mock the competitions of other reality shows. "Be on the lookout," says the voice-over. "Spa residents are frequent visitors here in Tomorrowland . . . to share their stories and to eat a square meal. The show is taped each week on this very stage to measure their progress."

The screen goes dark, and the feature presentation begins. It's called *The Human Ape.*

Dr. Sue Savage-Rumbaugh is a scientist who, in 1980, introduced a new approach for studying the potential for primates to communicate with humans. Unlike earlier researchers, she didn't try to teach her bonobo chimps to speak, because she knew their vocal hardware wasn't capable of producing human-like sounds. And she didn't try to teach them sign language either — as other researchers had done — because she knew their hands and faces lacked the subtle musculature to convey enough meaning. But most important, Dr. Sue didn't steal a baby chimp from the care of its mother, as *all* the other researchers had done. This was her first groundbreaking innovation. Kanzi grew up with his family intact, surrounded by the humans who were friends with his mother. He was immersed in spoken language, and he learned to point to symbols on a picture-board to facilitate communication. Kanzi and his younger sister Panbanisha have now, almost thirty years later, supplied ample evidence to support the claim that apes have the capacity to be people too.

Not all apes, however, deserve human status. Such a classification belongs only to those in such a group as Kanzi and Panbanisha's, who have had the benefit of humans to rear them. These bonobos have the language skills of a modern three-year-old human, and their play habits and emotions run parallel as well.

So, does this mean that bonobo chimps should be allowed to walk freely among humans? The film says *no*. Three-year-old human children aren't free to roam the world either, so why should an equivalent bonobo?

Equal rights follow a graduating scale, and the examples are everywhere. Children are not permitted to smoke cigarettes, drink alcohol, or vote in elections. And likewise, rogue governments are restricted from acquiring the advanced weapons of war.

The film ends with a closing statement: "At what point

does a creature qualify as human? It won't be long before this question becomes the great philosophical debate of the twenty-first century."

The crowd disperses, and I venture back toward the Grand Prix Raceway, where I recall seeing a restaurant. It's time for lunch.

"I'm giving your inheritance to the dog."   −Dad

CHAPTER 5

Cosmic Ray's Starlight Café

Outside the doors of the counter-service restaurant, known as Cosmic Ray's, there's a billboard showing the menu. Meals are divided into three categories:

> Counter 1 – Paleo
> Counter 2 – Primal
> Counter 3 – Keto

Above the billboard is a lighted screen that carries a traveling message. "Choose your style and place your order, then enter the restaurant and enjoy the show. Your number will appear when your food is ready for pick-up. . . . All meals are free of gluten, soy, hormones, pesticides, herbicides, antibiotics, preservatives, trans fats, artificial colors, artificial flavors, and ill-advised sweeteners. Nothing is irradiated or pasteurized, and all animal products come from naturally-grazed stock fed a minimum of grain. Please inform the staff if you have any other restrictions."

I order a Paleo meal of cold-water fish, steamed broccoli, and grilled vegetables, all of them smothered in avocado oil. For dessert, I switch over to the Keto menu and order a Chocolate-Chip-Cookie-Cheesecake Fat Bomb.

Inside the restaurant there's a video screen, where a cartoon Mickey Mouse introduces the next program. "No more

Mickey Mouse nutrition," he says. "Ha-ha. It's time to put the *credible* back into *in*credible and step outside the Disney domain." He moonwalks to the edge of the screen. "Sooooo, on with the show!"

Mickey disappears, and an episode of *Sesame Street* begins. The lesson for today is Basic Health.

Big Bird and the familiar cast perform skits and sing songs, offering cameo appearances to Mickey Mouse. This troupe delivers the following messages:

> Yes, sugar and grains have been the villains all along. This includes grain-derived oils.

> The first words out of every doctor's mouth should be: "Have you tried reducing your net carb intake to 30 percent of calories? Maybe you'd get better results at 20 percent. Ketogenic dieters — particularly those with epilepsy — shoot for 5 percent." For a 2,000 calorie day, such percentages equate to 150 grams, 100 grams, and 25 grams of net carbs, respectively.

> Of the three dietary macronutrient categories — fats, proteins, and carbohydrates — only carbohydrates are nonessential. The body can make its needed glucose from protein, lactate, and the glycerol backbone of fatty acids.

> Saturated fat is the body's preferred fuel, not glucose. Consider that the bloodstream carries only about a teaspoon of sugar and that the body must clear any excesses quickly. Not so for fat. In fact, glucose is converted into saturated fat for storage and later use, but not the other way around. Saturated fat is a

long-lasting, clean-burning source of steady energy. It should be consumed with gusto, not avoided.

The "keto reset" is a six-week dietary protocol for restoring the body to its factory settings, whereby cells are re-equipped to metabolize greater quantities of fat for energy. Once through this six weeks, fat-burning is easy.

Keeping insulin low is the secret to burning fat. Weight-loss diets fail when carb intake elevates insulin, trapping body fat and forcing the body to harvest its own muscle to make up the calorie deficit. The dieter loses weight, yes, but it's muscle instead of fat.

LDL cholesterol (pattern A) is not a menace. Heart disease is an illness of inflammation, not high cholesterol. Assess your real risk with the following two blood tests: *HS-CRP* and *NMR LipoProfile*.

To activate your body's advanced healing capabilities, do a three- to five-day fast perhaps twice per year. (Consult a knowledgeable physician.)

Over-training is the current standard for pursuing health and athletic performance. Ironically, better fitness and performance come from far less frequent trainings. The new mantra will soon be: "Maximum adaptive response; minimum wear and tear."

The best weight training lasts about 20 minutes and occurs no more often than once per week: five exercises, one set each, all the way to failure (in under

2.5 minutes, otherwise the weight is too light), super-slow reps (no counting), and never relaxing the load through locked limbs, resting positions, or setting the weights down. This method fatigues all four muscle fiber types before any of them can recover, sending a strong muscle-building signal to the body without imposing undue wear or risk of injury.

The best sprint interval training involves short bursts of all-out effort, lasting 5 to 15 seconds and separated by up to several minutes.

What is the body composition of the person giving you health advice? If it's excellent, then they might know what they're talking about. But if it's not, you can be almost certain that they don't.

    I pick up my meal at the counter and return to my seat, where the stage is now occupied by three *Sculpt Me Beautiful* contestants. They are sharing their stories of personal change, while also dispelling some of the common notions about dieting and alternative medicine.

"Think again, Bert. You're instructing people with cancer, obesity, and heart disease to eat more fat."
–Jane, B.S. Nutrition, RD

"Customized nutrition — it seems so obvious." –Tom

"Individuals and business owners must act on their own to improve health and their bottom line. Our food and wellness industries are so deeply invested in the status quo that they can't be relied upon to bring about change." –Bert

CHAPTER 6

Escape through Buzz Lightyear's Space Ranger Spin

TYPE: indoor adventure ride and shooting gallery
DURATION: 4½ minutes
CAPACITY: 2 to 3 riders per car
HEIGHT REQUIREMENT: none
LOADING SPEED: fast
AVERAGE WAIT: 3 minutes per 100 people ahead in line
RECOMMENDED AGES: 3 and up
DESCRIPTION: The Buzz Lightyear ride remains little changed
  by this retrofit. A vehicle, mounted with two laser guns
  and a joystick, travels through an arcade, but this arcade
  is now themed to be located within the hull of an ocean-
  going cruise ship.

I enter the pavilion, where a line of guests advances under
red emergency lights. The many Buzz Lightyear props and
figurines along the walls lie dark, activated only sporadically
by bursts of electricity.

"Attention, fellow passengers," comes an announcement
through a bullhorn. "You are the fortunate few who have
made it this far. Our ocean liner, *The U.S. Economy,* is in
trouble, and the only path to the lifeboats is through Buzz
Lightyear's arcade. We've redirected the track to get you
onto the deck.

"But don't be frightened. The situation is not as bad as

you've heard, and you'll know what I mean when you see for yourself the view from up top. While you're waiting, though, here's an explanation of what's been happening.

"We passengers and the crew boarded this ship having never before seen the ocean. We crossed the gangway over placid inland waters, then were locked below deck in this hotel without windows. Even the bridge is sealed off from daylight. The captain and his navigators have been piloting the ship by feel—keeping it steady by managing the throttle, moving the rudder, and adjusting the ballasts. They believe, as they were taught in school, that the high seas are naturally calm, but that squalls pose a constant threat.

"And the worst squall on record seems to be upon us, as platters have been reported spilled in the dining rooms, and as dancers have taken nasty tumbles in the ballrooms. Rumor has it that the ship is teetering on the edge of the earth . . . and the captain's reaction seems to confirm this. 'Reverse the engines! Crank the wheel!' he has cried. 'The earth is flat!'

"In economic terms, these flat-earth seafarers believe that *The U.S. Economy*, when sailing properly, should advance steadily year after year, and that fluctuations in unemployment, inflation, and consumer spending should be controllable by government intervention. So the government tinkers with what it can—interest rates, the money supply, the tax code, et cetera. But results tend to be mixed, as unwanted conditions sometimes persist or worsen.

"A few of us passengers, however, have found our way to the deck. We can see that the earth is not flat and that skies are not cloudy. In fact, the moonlit heavens reveal an endless expanse of rolling water that places ourselves, at this very moment, on the top of a giant swell. And, as everyone knows, each and every swell is followed by a trough. We watch with horror as our pilots turn the boat sideways and pump out the ballasts into the void. Our navigators are unwittingly trying to ride the crest of this wave forever. But the ship must slide into the trough — there's no stopping it. The important question is, in what condition will our boat reach the low point? Our current strategy threatens a capsize.

"Round-earth seafarers have studied the data and spotted the patterns. They know that an economy is driven largely by demographics — specifically, the age distribution of the population. It is well documented that people's habits for spending and saving vary predictably throughout their lifetimes. It is these habits that drive economic activity. Such demographic trends explain the Roaring '20s and the subsequent Great Depression. They explain the inflation of the 1970s and the booms of the '80s, '90s, and 2000s. They explain why the economy of Japan went from being the envy of the world in the 1980s to being mired in its 'lost decades' while the rest of the world prospered. And it is these demographic trends that long ago predicted the coming downturn, as the last of the Baby Boomers pass their peak spending years. There is no way to avoid the next Great Depression. The best that can be hoped is to navigate gracefully through it.

"The maneuver is a simple one, though it sounds suicidal to anyone below deck. 'Steer into the abyss!? That's insane!'

"But that is exactly what needs to be done.

"Every population bulge brings with it a booming stock market and a booming real estate market. It's an ever-recurring pattern. And each time, when the bulge finally passes, it leaves behind a credit crunch, as the collateral of mortgages loses value. There is no mystery here. The coming drop in consumer spending will drive down prices and wages throughout the economy. No amount of intervention can prevent it. Just study Japan for the proof. So, yes, the ship's correct heading should be directly into the trough, because that's exactly what it is—a trough and not the edge of the earth.

"Now, don't be surprised if when you reach the deck, you happen to hear laughter. Rest assured, it's not malicious in any way. The escaping passengers have finally realized that their pilots have little effect on the boat's course, and that all of the conspicuous boasting and finger-pointing among the crew has been a comedy of errors. For instance, the former captains, Ronald Reagan and Bill Clinton, deserve no credit for the rises that occurred while they were at the helm. Likewise, George W. Bush should not be blamed for the current downturn. And as for Captain Obama . . . so sorry, Mr. Limbaugh, there's no point in hoping for his failure; the new captain has zero chance of keeping the boat aloft indefinitely, because the demographics are entirely against him.

"When you get to the deck, try to find Howard Marks, as he's one of the few people who can secure for you a lifeboat."

The line reaches the Launch Bay, and I climb aboard a Space Cruiser. I move through the arcade and earn the rank of Space Cadet with my score. My vehicle then enters a dark passageway—identified as the Emergency Escape Hatch—where the arcade noises fade and are replaced by those of an ocean liner's cavernous hollows.

The tunnel brings me out onto the deck of a ship, and I'm greeted by the tropical air of a fine summer's evening. It's a commanding view from on top of this wave, featuring a panorama of other boats facing similar circumstances.

I step from my vehicle and follow alongside a railing that is overhung with lifeboats. On an opposite wall, instructional signs are posted for the captain and crew, should they happen to reach the deck in time:

1. Plug the leak of private money to public officials. *Campaign finance reform* must be the number one issue of future

elections until the ship's integrity is secured.

2. Perform the standard and well established *FDIC Intervention* on insolvent financial institutions one last time, regardless of size. Then end the fractional reserve system, which currently allows banks and insurance companies to gamble with the public's money. Instead, force them to become Limited Purpose Banks.

3. End the bailouts. The economy is better served when its failing businesses are purged. And please note that a bankrupt company does not necessarily shed more jobs than it would have otherwise. The fire-sale of a business merely puts buildings and employees on the auction block for new owners. The process is extremely unpleasant, and it deflates the economy swiftly, but getting to the bottom of the trough sooner means a surer and more efficient recovery, with less overall suffering.

4. Stop the stimulus packages. They won't work in any form, and they'll raise the national debt at a time when *Gross Domestic Product* is falling. The effect will be like that of doubling one's home mortgage while enduring a hefty pay cut. The *debt-to-GDP ratio* of the nation will soar to dangerous heights, just like Japan's already has.

5. Give financial assistance to the unemployed, but require them to work or go to school in exchange. Every dollar spent should be an investment in the future.

6. Issue *Green Cards* to college graduates who want to stay and work in the United States. This is precisely the demographic group needed to buoy the economy.

7. Keep the nation's borders open to commerce, and discourage the global trend toward protectionism. Remember what happened during the last Great Depression; closed economies are incubators for fascism.

8. Do the inevitable: raise taxes, slash government spend-
    ing, and let interest rates normalize. Be a role model to
    the world and begin this pain earlier rather than later.

"You're calling Warren Buffet a fool? He's *buying* companies
right now, not *selling* them."

> —Ferguson, B.S. Finance, CFA, CFP

"Most people are going to learn the lesson of demographics
the hard way. In economic terms, 2009 is 1929."   —Bert

"If history is indeed repeating itself, Bert, you'll enjoy the
following quotes."

> —Uncle Albert, Professor Emeritus, Sociology

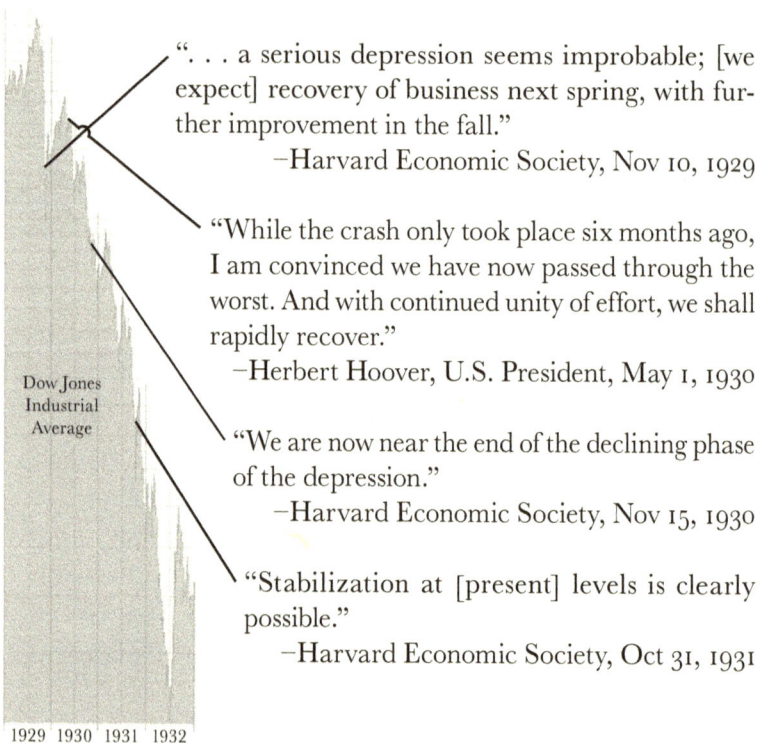

". . . a serious depression seems improbable; [we
expect] recovery of business next spring, with further improvement in the fall."

> —Harvard Economic Society, Nov 10, 1929

"While the crash only took place six months ago,
I am convinced we have now passed through the
worst. And with continued unity of effort, we shall
rapidly recover."

> —Herbert Hoover, U.S. President, May 1, 1930

Dow Jones
Industrial
Average

"We are now near the end of the declining phase
of the depression."

> —Harvard Economic Society, Nov 15, 1930

"Stabilization at [present] levels is clearly
possible."

> —Harvard Economic Society, Oct 31, 1931

1929 1930 1931 1932

CHAPTER 7

Walt Disney's Carousel of Progress w/Overlay

TYPE: audio-animatronic stage show and optional dark ride
DURATION: 20 minutes
CAPACITY: 226 guests per theater (6 theaters)
HEIGHT REQUIREMENT: none
LOADING SPEED: fast
AVERAGE WAIT: 5 minutes per 226 people ahead in line
RECOMMENDED AGES: all
DESCRIPTION: This Walt Disney creation from the 1964 World's Fair is a nostalgic journey through the life-altering innovations of the twentieth century. Six theaters revolve around six stages, stopping for five minutes at each one. The first and last stops accommodate loading and unloading, each in front of a giant video screen, while the other four stops each offer a stage show with robotic performers. A simultaneous dark ride runs its messages silently across overhead screens, placing the twentieth century in its proper historical context.

STAGE ONE: LOADING & PRE-SHOW

A door opens at the back of an empty theater, and I file in behind other guests to find a seat.

There's a video screen spanning the entire front of the stage, and it displays the ever-changing patterns of a kaleidoscope. Text soon overlays the image: "The *Product* Cycle."

The words linger for a moment, then dissolve into the background. A graph develops slowly in their place. The title is "Urban U.S. Households with an Automobile," and the chart twinkles at the point where its S-shaped curve passes through an inflection point. Labels soon materialize on the graph.

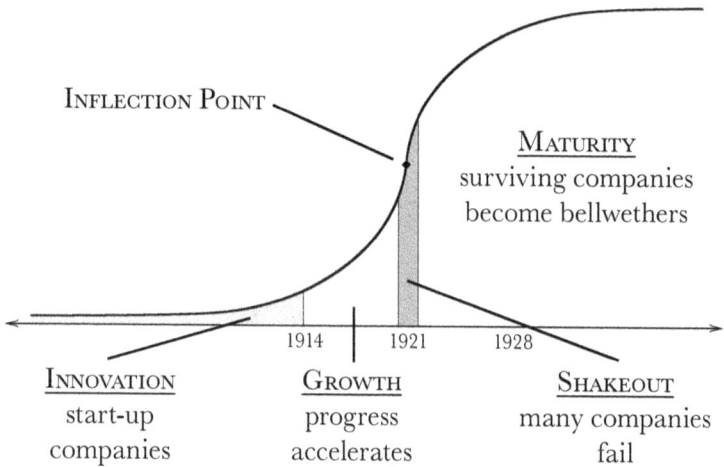

The graph shrinks away to the side and is replaced by a new one: "U.S. Households with a Radio." The only differences between this and the previous chart are the labels on the horizontal axis. The years now read 1922, 1932, and 1942.

This second graph, too, moves sideways to make room for yet another: "U.S. Households with a Television" — 1950, 1955, 1960.

The graphs now come in rapid succession: "U.S. Households with a VCR" — 1984, 1987, 1990; "U.S. Households with Internet Access" — 1993, 2000, 2007; "U.S. Households with a Cell Phone" — 1994, 2001, 2008; "U.S. Households with Broadband" — 2000, 2004, 2008.

Each curve develops slowly at first, then accelerates

through its inflection point before slowing to the upper right.

The graphs soon come too quickly to read, and they fill the screen to capacity before abruptly hiding themselves behind the first three graphs along the left edge.

A new headline appears at the center of the screen: "The *Economy* Cycle." Once again, a similar chart develops.

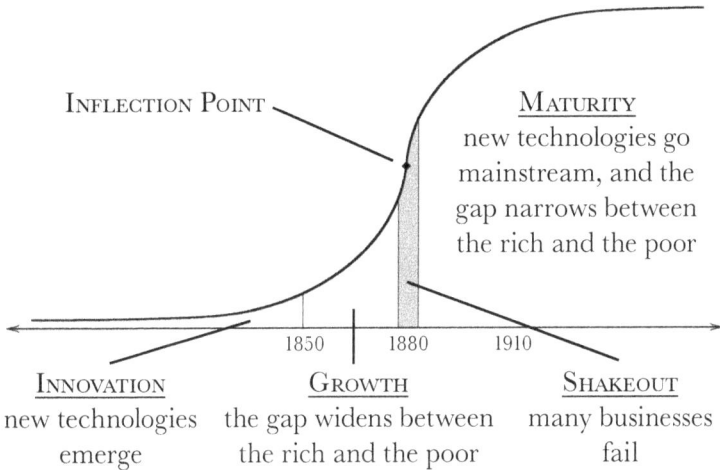

INFLECTION POINT

MATURITY
new technologies go mainstream, and the gap narrows between the rich and the poor

1850   1880   1910

INNOVATION
new technologies emerge

GROWTH
the gap widens between the rich and the poor

SHAKEOUT
many businesses fail

Upon its completion, the graph slides upward to the right, and a new graph replaces it that is identical except for the years: 1905, 1935, 1965. It, too, moves to the right to occupy the edge of the screen.

Another graph appears — this one having years that extend into the future: 1985, 2015, 2045. This third graph shifts to the lower right corner, leaving the center of the screen open again.

An unseen announcer welcomes visitors to Walt Disney's Carousel of Progress and begins an introduction of the ride.

Behind his narration, the words "The *Progress* Cycle" appear on the screen. A new graph plots its S-curve extremely slowly, moving from left to right past the year 10,000 BC. The line seems to be advancing at the rate of a few centuries per second.

By the time the line reaches 1450 AD, it is rapidly accelerating upward, and when it hits its inflection point, the whole screen flashes bright white, then turns completely dark.

A message scrolls across the top of the screen:

> The moment of steepest ascent is called the *Singularity*. . . . Humankind is just a few decades away from it. . . . The *Growth* phase has ended. . . . Few worlds survive the *Shakeout*.

The joyous strumming of a banjo marks the end of the narrator's presentation, and a jubilant choir sings the song *There's a Great Big Beautiful Tomorrow*. The text on the screen freezes in place, as the theater begins its rotation to the right.

STAGE TWO: ACT I
The male performer is a life-like robot who sings along with the choir. He sits on a rocking chair in his 1890s kitchen, holding a newspaper in one hand and a pipe in the other. He goes about explaining what life was like at the start of the twentieth century.

A video screen is fastened to the ceiling above the stage, suspended like a ribbon across the front of the theater. It initiates a streaming line of text:

> The first markers of progress were separated by hundreds of millions of years: . . . 3.5 billion years ago, microbial life . . . 2.0 billion years ago, single-cell life . . . 1.2 billion years ago, multi-cell life . . . 0.6 billion years ago, simple animals.

Then! . . . at about 530 million years ago — the Cambrian Explosion! . . . The markers of progress grew closer together by a factor of ten: . . . 500 million years ago, fish . . . 360 million years ago, amphibians . . . 300 million years ago, reptiles . . . 200 million years ago, mammals . . . 60 million years ago, primates.

As you can see, the advance has been accelerating: . . . 7 million years ago, humanoids . . . 2.5 million years ago, stone tools . . . 1.5 million years ago, domesticated fire . . . 150,000 years ago, the first modern humans . . . 25,000 BC, ceramic art . . . 15,000 BC, domesticated animals.

Progress does not follow a straight line; . . . it grows exponentially.

And, speaking of progress . . .

These final words appear on the screen just ahead of the character on stage saying them aloud. The banjo returns, and the performer rejoins the song *There's a Great Big Beautiful Tomorrow*. The text, once again, freezes overhead, as the theater begins advancing to the next stage.

STAGE THREE: ACT II
The same character, in different clothing, sits in a kitchen that's been updated to the 1920s. His musical accompaniment has been updated too, performed by a jazz combo.

As the song finishes, and the theater comes to a halt, the text begins flowing across this next stage's ceiling-mounted screen:

At 10,000 BC, a few of the human tribes settled into villages. Their progress was astounding: . . . 9000 BC, agriculture and pottery . . . 8000 BC, copper and the plough . . . 7000 BC, animal husbandry . . . 6000 BC, proto-writing . . . 5000 BC, irrigation and beer . . . 4000 BC, the wheel and axle, and more beer . . . 3300 BC, bronze . . . 3000 BC, sailing . . . 2000 BC, currency . . . 1500 BC, iron.

And then came self-awareness! . . . It made its debut around 1000 BC. . . . Self-awareness put the power of willful reflection and contemplation into the minds of its human hosts. . . . Progress went from being separated by millennia to being separated by centuries.

The list quickly becomes exhausting, so here are just a few of the highlights, all from China: . . . 200 AD, paper . . . 700 AD, printing . . . 900 AD, gunpowder.

The next pivotal year was 1440 AD, when Johannes Gutenberg introduced the movable type printing press to Europe, where the region's languages used smaller, more practical character sets than those of neighboring Asia. . . . Suddenly, the reflections and contemplations of fellow humans could be shared like never before. . . . The time scale of progress shrank further. . . . By the dawn of the twentieth century, the rate of advance was generating significant markers less than twenty years apart.

It just goes to show that . . .

These last printed words are also spoken by the character

on stage just as the music begins, and I listen to another round of *There's a Great Big Beautiful Tomorrow.*

STAGE FOUR: ACT III
The new kitchen is from the 1940s, and the jazz combo has been replaced by a swing-era orchestra. The usual host, named John, belts out the final verse, and I turn my eyes upward toward the overhead screen. The theater comes to a stop:

> It's not surprising that Walt Disney was fascinated by progress; his generation experienced more of it than any other in human history.

> In earlier times, the rate of technological advance was so slow that it could barely be felt. . . . In fact, the word "progress" didn't acquire its current meaning until the year 1610, during the period when the turn-over-rate of change fell to within one living memory.

> But now, the rate of change is even faster, posting several doublings per lifetime. . . . The stories of your parents will pale in comparison to those of your children.

> Knowledge is the great driver of progress. . . . And with each accrual of new knowledge, so too do the tools for gathering it and spreading it compound in their powers. The entire enterprise is self-propelling. . . . By 2050, the doubling of progress may be measured in months or weeks, or possibly even days. . . . Can you imagine?

So why are the heavens so silent to our instruments
of detection when we search for other intelligent life
in the universe? . . . The nature of progress suggests
the answer.

John asks the audience to sing along with him, and the
theater starts to turn.

STAGE FIVE: ACT IV
John and his family reside within the open architecture of a
modern retro-future home. The song and theater come to
a stop, and words begin streaming across another screen:

The *Shakeout* is the period of instability that begins
before the *Singularity* (the inflection point of the
curve). . . . Such a period began here on earth in
the 1950s with the proliferation of atomic weapons.
. . . So far, the human race has successfully dodged
its own destruction. . . . But such an ambition will
become ever more difficult as time presses on.

The *Singularity,* which is slated to arrive by mid-
century, will usher a period of such radical instability
that a worldwide holocaust is highly probable. . . .
The powerful new technologies set to converge are
genetics, robotics, nano-devices, and artificial
intelligence. . . . They promise to solve the world's
problems, if they don't destroy it first.

The *Shakeout* will proceed at full force through the
Singularity and beyond. . . . And during this time, life
will hang by a thread over the jaws of disaster. . . .
Just one hateful act . . . just one careless accident . . .

just one mere oversight by the best and brightest of minds . . . could unleash a terror so vivid that no science fiction thriller could fairly depict it.

The stakes are high and getting higher. . . . The twentieth century will soon be remembered as a period of blissful innocence.

The song erupts with its optimistic words, "There's a great big beautiful tomorrow, shining at the end of every day." And the theater turns.

STAGE SIX: POST-SHOW & UNLOADING
The final stage is just like the first one: covered by an enormous video screen displaying a kaleidoscope. The graph at the center is labeled "The *Progress* Cycle," and it restarts its race toward the inflection point. This time, however, the white flash of the Singularity does not wash out the image. Rather, the S-curve continues onward and upward.

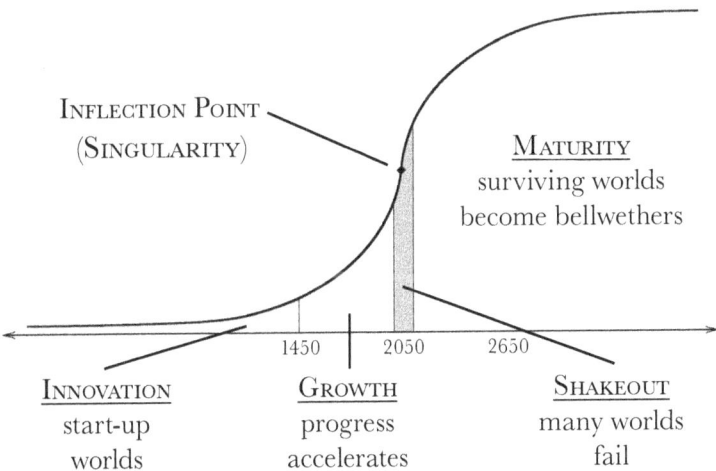

INFLECTION POINT
(SINGULARITY)

MATURITY
surviving worlds
become bellwethers

1450   2050   2650

INNOVATION
start-up
worlds

GROWTH
progress
accelerates

SHAKEOUT
many worlds
fail

Text appears below the graph:

> To survive the *Shakeout* means a joyous exploration
> of our great big beautiful universe.

The announcer instructs the audience to collect their
belongings and exit through the rear doors of the theater.
"And remember," he says, "tomorrow is just a dream away."

"The *S-curve of progress* reveals that our technological advance
is rocketing toward a flashpoint of alarming and breathtaking
change. If we don't do our best to prepare for it, we're surely
in for trouble." —Bert

"Your glimpse beyond the Singularity will no doubt get the
pundits talking."
                —Tom, B.S. Physics, M.S. Atmospheric Science

"I wish we had known about the S-curves earlier, Bert. We
could have made a fortune during the Tech Wreck." —Mary

CHAPTER 8

Nano Joe's Coffee Shop
(formerly *Stitch's Great Escape*)

TYPE: theater-in-the-round science fiction show
DURATION: 12 minutes (plus 6 minutes of pre-show)
CAPACITY: 162 guests per theater (2 theaters)
HEIGHT REQUIREMENT: none
LOADING SPEED: moderate
AVERAGE WAIT: 6 minutes per 162 people ahead in line
RECOMMENDED AGES: 3 and up
DESCRIPTION: This attraction is a kid-friendly rework of the props left behind by *Stitch's Great Escape,* meaning the shoulder restraints have been removed and the scare factor has been reduced to near zero. The performers are audio-animatronic, and the humor is "bathroom." Two rooms of pre-show prepare the audience for the main presentation.

ROOM ONE OF THE PRE-SHOW
The line enters a spacious foyer, where large glass tubes, filled with coffee beans, obstruct the view into the café. I can hear patrons and espresso machines beyond them, and the air is rich with the fragrances of a hundred fresh brews.

"Welcome, coffee-lovers," says a male voice from a speaker system. "Thank you for visiting Nano Joe's — where tomorrow's best coffees are served today! While you wait for your turn at the counter, please allow me to share with you the history of this fine establishment. You see, this building used to house the Nanotech Institute—a thriving research facility dedicated to the development of nano-sized devices. Now, for those of you who may not be familiar with the term, a nano-device is a machine so small that one-hundred thousand of them could fit across the end of a single human hair."

A video screen on the wall shows a short animation to demonstrate the minuscule size of a nano-particle.

"But the results from this very lab proved to be the death knell for the industry. More and more of our studies were concluding that nano-particles were finding their way into the most sacred of places—namely, the cells of living things. These precocious little buggers were slipping across cell boundaries and mucking up the works. Nanotechnology was deemed too dangerous to pursue, so a worldwide ban went into place. This Nanotech Institute was just one of dozens of labs around the world that had to be shuttered."

A chime rings. "DNA scan complete," says a woman's voice. "Comparing to registry."

"But stopping the research," the narrator continues, "was the most dangerous decision of all. Instead of simply banning the release of nano-materials into the environment, the world's governments imposed a sweeping prohibition against any further development of the technology."

Another chime rings. "Registry check complete," says the voice. "Law enforcement detected, but outside jurisdiction. Cleared for entry."

"Good news for *you*," says the narrator.

A hidden door in the wall slides open, revealing a passageway that echoes with parlor music.

"Nano Prohibition ends here. Welcome to the twenty-first-century speakeasy. I trust you'll keep this under your hat. But no matter; our search-and-destroy nanobots are fully capable. Please proceed to the Secret Lobby."

ROOM TWO OF THE PRE-SHOW

The Secret Lobby is a dark room that features a metallic robot set between two empty glass tubes, both of which glow brightly along a wall.

"Come in! Come in!" says the robot, beckoning us with his arm. "There's room over here!" He points. "Be sure to hold up the young ones so they can see me!"

The robot places his hands over his face and plays peek-a-boo into the crowd. He then extends both arms and waves.

"Hi there," he says. "My name's Meterbot. What's yours?"

Two children from the audience shout out their names.

"Nice to meet you," he says.

The robot scans the crowd. "Show of hands, please. How many kids here are glad they're not in the coffee shop right now, listening to the grown-ups go blah, blah, blah? Believe me, I know the boredom . . . *and* the bad breath!"

Meterbot tips his nose into the air. "But I can smell it anyway," he says. "I have a super-sensitive olfactory, and I can tell when people have been eating the wrong foods. Usually it's coffee or pop or bread or milk, but sometimes even citrus fruit can cause a vile odor to seep out if it's in the wrong body type. So, to all of you kids who might be wearing a dirty diaper right now, take heart. At least your diaper region is *supposed* to stink.

"Well, I guess I should get down to business now and prepare you for the main event. You'll be entering the Nanotech Theater, where our goal today is to show you that research and development can't really be stopped. The unintended consequence of prohibition is to drive an unwanted activity underground. Which raises the critical question: who do you want steering your progress? —the law-abiding professors, or the hacks and malcontents of society? Here at the Nanotech Speakeasy, this secret underworld is in the capable hands of the freakish . . . the outrageous . . . the notorious stooge of the stage: Nano Joe. He's a renegade researcher who currently earns his living as a stand-up comedian.

"Now," says Meterbot, "are there any teenage girls in the house?" He points around as if counting them. "Ah, yes, and some teenage boys, too. You're going to like what you see today. Nano Joe is a middle-aged man by the calendar, but his physique is that of a nineteen-year-old gymnast. He's got the body you've always wanted for yourself, in one way or another.

"And to you kids of every age, please raise your hands and wiggle your fingers. Are they working okay? . . . Great! . . . Now, reach under the waistband of someone you love and grab for their underwear. And when you think you've got hold of it, need I say, pull."

Laughter and little chirps erupt from around the room, as families and friends wrestle each other.

"Victims!" shouts Meterbot, "don't just stand there taking it! Go ahead and tickle them while their hands are busy. That's the way. I love this game; it really fills the cracks."

The music grows louder, and the theater doors swing open.

"Bye, now," says Meterbot. "Enjoy the show. Hands to yourselves, please. Sanitizer on your right. Barf bags on your left. Anyone need a throat lozenge? Please enter and pick your seat."

THE SHOW

This theater-in-the-round encircles a small stage that is centered by a single glass tube. Inside the tube stands a human figure that looks to be roughly nineteen years old. He's talking on a cell phone with a young woman, whose face and torso appear on the many flatscreen monitors around the room.

"What do you think of my new look?" she says. The woman presses a button on her wrist, widening the picture to reveal her whole body. Her figure is sleek and well toned, and barely covered by a bikini.

"Nice," says the man in the tube, "but keepin' it G here; I'm workin' right now." He pushes a button on his own wrist and returns the image to the close-up. "I'll catch you tonight——"

"——at the Masquerade Club?"

"Where else? . . . Love ya. Ba."

He punches more buttons and rewinds the video of his phone call back to the bikini shot. He leaves it fixed on the monitors for all to see.

"Don't be shy," he finally acknowledges his audience, which has almost finished seating itself. He crosses his arms and pushes up his biceps under the tight sleeves of his fluorescent-yellow t-shirt. He then raises one eyebrow to the crowd.

"I'm Nano Joe," he says. "I bet you can't guess why I'm sealed inside this tube. . . . Or maybe you can. . . . Meterbot warned me there's an odiferous contingent among your *ranks*." He laughs to himself.

"Now, I'm hoping you'll be willing to share some information with me," he says, and he hooks his thumbs through the belt loops of his jeans. "Please raise your hand if you've eaten any food today. . . . Did you brush your teeth afterward? . . . Are you wearing any deodorant right now? Is it working? . . . How about your neighbor's? . . . And what about deodorizing socks? No checking, please. . . . Anyone wearing sunscreen or make-up?

"Welcome to the world of nano-particles. Each of the things I've listed—things which, for decades, have performed just fine without them—can now bring you into contact with nano-materials. But don't go searching the labels of your food and shampoo for the word 'nano' if you're trying to avoid them. You'll have to do more digging than that. Disclosure is currently not required."

Nano Joe places a finger over his left wrist.

"Anyone want to venture a guess at the real reason I'm standing inside this tube?" A loud release of flatulence overlaps his final words. "And that's not it," he snickers.

Two robot arms, which extend down from the ceiling, spring to life and bend away from the glass tube, as if escaping a noxious fume.

"Oh, stop it," says Nano Joe. "You're such ninnies."

The two arms turn sheepishly back toward him.

"I'm sealed in a tube," he says to the arms, "and we do this bit a hundred times a day. I thought you'd've caught on by now."

Nano Joe fans the air behind the seat of his pants. "Ahh," he sighs, "it's colorless and odorless . . . just a sound gag, really. And there's a lot more where that came from. I get a full day of expulsions from a single recharge . . . and all at the touch of a button." He points to his wrist.

"But enough about me. What about you?"

He turns his eyes upward. "Armatures! Do your business and sniff out the guests."

The two robot arms launch into action. They rotate at the ceiling and bend at their elbows, spinning their end-mounted sensors like the barrels of a Gatling gun. Each armature eventually settles its aim into a different region of the audience.

"Good work, boys," says Nano Joe. "It seems two of our grown-ups here have developed a mild case of the Rancids. Let's give 'em a good scrub-down, shall we?"

The lights go off in the theater, replaced by a single strobe. The giggles of a man and a woman can be plainly heard within the room, though it's too dark to find them and see what's actually happening to them. The one notable visual effect, however, is that Nano Joe's t-shirt has somehow disappeared. His upper body is well defined and muscular, and he poses like a body-builder for the audience.

The giggles subside, and the house lights return. Miraculously, Nano Joe's t-shirt is back on.

"Did you enjoy the strobe?" he says. "It penetrates the fibers of my shirt. I encourage you to purchase one in the gift shop and wear it to the next show."

Nano Joe's phone rings, and he raises his hand. "I'm taking this call," he says, then he switches off his microphone.

The robot arms at the ceiling take over the entertainment, moving slyly around Joe and performing silly gestures behind his back. They telecast their antics over select monitors around the room, quickly reverting the screens to the picture of his girlfriend whenever he looks around. The armatures take practice swings at his head — one pretending to hold a golf club and the other pretending to hold a baseball bat — until Joe finally ends his call and reactivates his mic.

"It's time for the demonstrations," he says.

A fog develops around him in the glass tube, creeping down from above through an eerie red light.

"This tube is the *growing chamber* for the nano-devices," he says. "My body can withstand them because it's less biological than yours. But don't worry, you're not that far behind me. All of us are well on our way to transcending our own biologies, even though some of us aren't ready to admit it yet. Take my word, the tampering's been going on for ages. Just ask Captain Hook to show you his left hand the next time you see him."

Nano Joe puts a hose into his mouth and draws a deep breath. "This air is just for talking. I've replaced my red blood cells with nanobots that carry oxygen far more efficiently. I could jog for fifteen minutes on a single breath if I wanted to. And check out my arm." He holds up his left wrist. "My *personal digital assistant* runs on blood sugar. No more batteries for *me*. Face it folks, we're fast becoming cyborgs."

The light source over his head switches from red to green.

"Ah," he says, "everything's ready." He pushes a button, and the mist vanishes from the tube. An aroma of chocolate fills the theater.

"That's much better, isn't it?" he says.

A second wave of mist develops in the tank, and again, the light changes from red to green. A new odor is released, this one reminiscent of an old-time candy store.

"One more," says Nano Joe, and he appears to be scrolling through a menu on his wrist device. "Not *pizza*. Not *petunia*. . . . Oops! I almost hit *pathogen*. I really should get that one off the list."

Nano Joe's phone rings again. "I'm taking this one too," he says. "Armatures, do your thing!"

The robot arms nod confidently and cue some music. Together, they perform a choreographed dance routine to a synchronized light show. Nano Joe grooves along with them as he converses, flexing his muscles whenever the strobe light comes on.

A dramatic finale brings the musical number to an end, and the house lights return. The robot arms appear to be panting, as if out of breath, and they stare at Nano Joe, who is still talking on the phone.

One of the armatures taps on the glass tube, startling Nano Joe and prompting him to end his call immediately.

"And for my last trick," he says, "here's my latest prototype invention—Tinker Bell. She's a self-replicating nanobot that delivers targeted doses of pixie dust. I guarantee you'll feel the Disney Magic."

A gold mist surrounds Nano Joe within the tube.

"Each of these teams of flying Tinkerbots is assigned a seat number. Every one of you will get a moist spritzer of pure delight and feel the breeze of a thousand tiny wings."

The house lights go off, and the red light above him turns to green. The miniature Tinker Bells glow like fireflies and escape through the top of the tube. They spread out across the ceiling, swirling and dancing overhead, eventually dropping their promised payloads of damp pixie dust. A minute later, they've all swarmed back into the tube.

"These nanobots are pre-programmed and quite sophisticated . . . far more advanced than the earlier odorbots. The Tinkerbots return for recycling, and each and every one of them is accounted for. The last thing you want is a self-replicating nanobot out on the loose. And as for the odorbots, they've mostly drifted to the floor by now. We'll be sweeping them up after you leave. Yeah, right."

A throbbing drumbeat marks the return of the dance music, and the robot arms restart their hip-hop routine.

"And hey," shouts Joe over the music, "don't worry about the particles you've inhaled here today. Nobody cares enough to measure your exposure. Ha! Ain't life a kick!"

He joins in on the dancing, and the theater doors come open.

My exit passes through a gift shop, where the background music features voices that sing of *miracles from molecules* in the style of an old show tune.

Mannequins of exaggerated human form are on display along shelves near the ceiling. Their body styles range from conventional to exotic, and nearly all of them boast of being compatible with a bizarre array of power-tool attachments. All of them use *audio signals* rather than *pain* as their primary indicators for damage. And *knowledge* and *skill sets* are available as independent downloads. Snap-in processors offer "more thoughts per second."

I browse the racks of clothing but am unable to find one of Joe's disappearing t-shirts. Instead, I settle for a sweatshirt that reads, *Child of the Singularity — Young Enough to Live Forever.*

"Human-made nano-particles could ruin the food chain from the bottom up, or perhaps disrupt the biosphere in another catastrophic way. Our wait-and-see habit of letting damages accrue before enacting regulations might actually fail us this time. Nano-particles need to be accounted for and recycled. The advance of technology is narrowing our margin for error." –Bert

"So tell me, Bert. Do you want people to trust their experts or not? You praise them in one breath and slay them in the next." –Jane

"I've got a joke. What's the difference between an expert and a stand-up comedian? Answer: The comedian generates laughter on purpose." –Uncle Albert

CHAPTER 9

Space Mountain *Backups*

TYPE: action-adventure on a VRC (Virtual Reality Coaster)
DURATION: 20 minutes
CAPACITY: two riders per pair of harnesses
HEIGHT REQUIREMENT: at least 52"
LOADING SPEED: fast
AVERAGE WAIT: $2\frac{1}{2}$ minutes per 100 people ahead in line
RECOMMENDED AGES: 13 and up
DESCRIPTION: This near full-immersion virtual reality simu-
lator is housed within Space Mountain's iconic structure.
Computer-controlled harnesses suspend guests below
a roller coaster track fitted with motors and brakes to
supplement the freewheeling. Riders wear headphones
and stereoscopic video goggles, and their hands operate
controllers.

The line begins in Space Mountain's entrance corridor, where
a speaker system delivers the following broadcast.

> Welcome to Space Mountain Backups—your one-
> stop shop for scanning, storage, reconstitution, and
> ultimate-risk adventure. That's right. Treat yourself
> to an all-or-nothing thrill ride. There's no better time
> to crash than right after a backup.
> Space Mountain's new partnership with the

United Nations Peacekeeping Force and the Milky Way Alliance means that all of our services now come to you free of charge. Just two mercenary adventures is all it takes to completely eliminate your cost. But there's more. Space Mountain captures your action on film, whatever the outcome, so you can enjoy your escapades again and again. It doesn't matter how you emerge—an original or a reconstitution—you won't miss a moment of your life.

The hallway is lined with large windows, offering peeks into the many reconstitution rooms of Space Mountain. Each room has a track along the ceiling that extends from a doorway to a stainless steel vault. In a few of the rooms, an empty harness travels to one of these vaults. But in most, the harness is occupied and sits parked, facing away from the windows, with its rider watching a violent video that always ends in darkness.

And don't worry about the pain and suffering of an agonizing death. All of our harnesses come equipped with high doses of morphine, so you'll get an instant injection upon any impact or extreme temperature swings. Dying hard never felt so good.

Space Mountain offers the latest in scanning and storage technology, with ten-times error-checking, secured encryption, and multiple-site holding. Your singular identity is guaranteed. And recovery is a snap. Our state-of-the-art Reconstitution Center restores your exact parameters to those of your most recent save.

Thank you for choosing Space Mountain Backups . . . and let the adventures begin.

The line enters a vast room that vaguely resembles a meat-packing plant. Tracks along the ceiling deliver hanging pairs of harnesses to multiple loading stations, where cast members help passengers to board. Once guests are locked into place, they are fitted with electronic headgear and rolled forward into a queue, headed for the Backup Chamber.

Each stretcher-like harness features a small platform to stand on and a cushioned backrest to lean against. Padded restraints pull down over shoulders and chest, and steering levers offer a place to rest the hands.

When my turn arrives, I step onto a hanger and pull down the bars. The steering wheel has a button under each of my thumbs.

Across from the loading stations are interior windows, where harnesses stream past in a shadowy flow. I assume these are the riders on their way to their second adventure.

The cast member helps me to strap on my goggles, then she lowers the headphones over my ears. The goggles are dark and burnt-looking, and the headphones hum with static. My harness rolls forward and sits for a moment, then engages with the track's motors.

Miniature sponges and squeegees appear on the surface of my goggles and wash the lenses clean. I'm traveling in the queue, bound for the entrance to the Backup Chamber.

As soon as I cross its threshold, the second harness of my pair detaches and moves forward to create a single file line. I peek out from under my goggles to verify that the harnesses have not truly separated. In fact, I can see a long procession of paired harnesses moving quietly through the Space Mountain interior. I refit my goggles and return to the virtual reality experience.

The Backup Chamber is a small room that ends in a

downward-sloping tube. Each of the harnesses ahead of me tips backward to a forty-five degree angle, then enters the tube, where it bathes in a bright light.

The room hums louder as my harness nears the tube and begins to recline.

"Breathe deeply and relax," says a female voice into my ears. The static is now gone. "Body-scan and neural-mapping commences in three . . . two . . . one . . ."

My harness trembles, and I shut my eyes against the glare. I can feel myself descending. . . . Then, everything becomes quiet, dark, and still.

I open my eyes and discover myself traveling from a stainless steel vault to a position in front of a television monitor. I'm in one of the reconstitution rooms.

"Welcome back," says a friendly voice. "Congratulations on completing your first adventure. Needless to say, you didn't survive it. But here are the highlights we've saved for you."

The screen on the wall plays an edited video, showing a battle scene between numerous iceboats, all of them racing across a frozen tundra. My mission, I'm told, was to destroy a terrorist cell that was plotting an attack against a nuclear depository in Canada. The rock group Metallica supplies the sound track.

Most of the film footage is from the perspective of one particular boat — no doubt mine — cruising at a tremendous speed. I am apparently the gunner up front, and I wield a laser weapon that fires a continuous beam. I watch as my former self burns away the running-gear of enemy boats and creates holes in the ice for them to crash into. Helicopters rally to the enemy's defense, and I'm subjected to gunfire and falling explosives. I torch a few of the choppers from the sky, then my video screen goes dark.

"It's good to have you back," returns the voice. "You're

fully restored and ready for your second adventure. Let's get you started."

My harness springs into motion and follows the track out the door. It joins a parade of other single harnesses and files past the windows overlooking the boarding area.

I bypass the Backup Chamber and merge with its outflow, accelerating into a tunnel of blue chaser-lights. Groups of harnesses peel off, left and right, into branching tunnels that connect to it.

My harness and three others turn right and travel a short distance into an underground airplane hanger. A small craft awaits us, and our harnesses are fastened to the rear wall of the cockpit. The pilot studies us beneath the rim of his safari hat.

"I'll be your shuttle service today," he says. "— Name's Thal Amus." He flips a switch, and the plane rises through a dark elevator shaft. "I'll have you to the battle zone in a jiffy."

Glimmers of daylight eventually stream through the windshield, and our plane emerges into an open-air temple at the top of a stone structure in Adventureland. Our lofty position provides us with a great view of the Cinderella Castle.

Thal throttles the engines, and we float free of the structure like a Harrier jet. He pulls another lever, and the engines redirect their thrust, shooting us forward out over Liberty Square. We enter a steep climb and head into outer space.

The acceleration continues, pressing me firmly into my backrest. Scattered starlight fades to blue, then vanishes.

A sudden deceleration returns the star fields, and our craft has arrived within a convoy of other space ships.

"You'll be on the biggest one," says Thal. "Be sure to note the look of the X-wing Fighters; you don't want to shoot the good guys. Your job is to protect them while they attack the Death Star."

Thal flies toward the largest ship in the fleet and carefully steers us into its landing port. He remains parked there while we, in our harnesses, are carted away on dollies.

I'm rolled at top speed through a labyrinth of hallways and into a turret, where I'm locked into place. A recorded message instructs me on the use of my swivel-action turret and dual cannons. A heads-up display promises to forewarn me of approaching vehicles.

Off in the distance, a dark moon-sized planet comes into view, and my accompanying fleet spreads into an arc around it. X-wing Fighters spill from the Alliance's many ships.

The first wave of enemy fighters washes onto the scene, and the battle begins.

The Alliance flyers are sorely outnumbered, and they swing past my position for relief, usually with two or three of the enemy on their tails. But the enemy fighters are no match for my armored turret and powerful gun.

News arrives over the intercom, that kamikaze pilots have destroyed the ship's bridge. We're on a collision course with the Death Star, and it's impossible to prevent. "Abandon ship," comes the order.

I continue firing my cannons. There's no way for me to disengage from the fight and flee; it's not in my contract. I'm locked into this turret to do or die. The loss of my harness will be the proof of my demise, forming the legal basis upon which Space Mountain can reconstitute me. I should strive to make my final transmissions glorious ones so that I can be proud of the film I'll reawaken to.

The dogfight continues outside, and three enemy fighters take a run at me. I turn my gun on them and destroy each one in turn, but their debris is destined to slam into me.

An instant later, everything goes dark.

"This one's alive, sir," I hear a voice. Then the commotion of the spaceship returns. "—Splinted and ready to go."

"To the escape pods, then. There's no time to lose."

I'm wheeled through the ship and sealed into a small compartment. A quick burst, and I'm jettisoned into space.

Through my tiny window, I see hundreds of other pods, like my own. We are all drifting away from the mother ship, as she receives a pounding of laser blasts from the doomed Death Star. A few of the ship's turrets still fire in her defense, but it's not long before they, too, fall silent.

The fleet is on the run now, and the last of the fighters has zoomed past our helpless pods. I glance back at the Death Star and bear witness to the fiery fissures now spreading across its surface, as the mother ship continues to penetrate it.

Pow! The Death Star explodes.

A wave of heat rushes through the sea of pods, followed by a shower of scorching shrapnel. Some of the pods are destroyed outright, while others are knocked off course. Mine is sent spinning.

Again, my surroundings eventually settle, becoming dark and quiet. I seem to be lying down with my eyes closed and traveling through a hallway.

"What's the story?" I hear a woman's voice.

"They don't know how it happened," replies a man. "The containment harness must have snapped open without signaling the breach."

"How unfortunate," says the woman.

"They couldn't identify this one as ours without the harness, so the Alliance is claiming no fault."

"I suppose the backup's already been made and set free."

"Yeah . . . just left the unloading area."

"Take care of it then," she says. "Get both of 'em back on

adventures until one of them buys it. We can't very well have two of them walking around, can we?"

"But this one's unconscious," says the man, "—hardly a sporting chance."

The two of them laugh.

My harness tips up, and my vision gradually returns. I'm fastened to a track along the ceiling and moving through a hallway. Eventually, I pass in front of the windows that overlook the boarding area, and I'm shocked by what I see.

There, being loaded into a harness and locked into place is the reconstituted copy of myself. I know it's really just the video playback of my earlier boarding of the ride, but in this world of virtual reality, it's the duplicate of myself who thinks that both of today's adventures ended in his death. He doesn't know that I actually survived the second adventure, meaning he shouldn't have been created. Yet, there's no way for me to apprise him of the situation. And even if I could, I know he'd argue for his own survival. He, like me, wouldn't volunteer to be the one terminated.

My clone glances in my direction, but I know he can't see me in this darkened hallway.

I zoom into the blue tunnel and race ever faster toward my third adventure—one I didn't explicitly grant permission for, although I probably actually did somewhere in the fine print. My harness splits off to the left and travels for a while, then slows, eventually coming to a stop in complete darkness.

The air is warm and smells of seawater, and I'm bobbing up and down.

An overhead door swings open in front of me, and I can see that I'm at the back of a boat garage. A shirtless man starts the engine of a speed boat and guns it. His female coworker wears a swimsuit and lets out some tow rope. The rope is connected to the lower portion of my harness, and it eventually draws tight, yanking me forward across the water like a skier.

I exit the garage and see thirty or so other setups like my own. Each of us Space Mountain mercenaries is in a harness equipped with a motorized parasail, a mounted gun, and a canister behind the backrest.

A simulcast of James Bond music blares from the boats, and the women riders recline on the decks, applying suntan oil to themselves. We are headed out to sea.

The boats pick up speed, and our sails eventually catch air, lifting us off the water. The towlines lengthen to several hundred feet, and each of us settles at a different cruising altitude. Mine is one of the highest.

"We'll be surrounding a volcanic island," comes a woman's voice through my onboard speaker. "When I release your tether, your parasail will automatically fly you directly over the crater. Your job is to use the left button and release your bomb *into* that crater. The steering and right trigger is for aiming and shooting your gun. Good luck. You'll need it."

Our fleet of speed boats forms a wide circle around a tiny tropical island, and the propellers of our parasails begin operation. The boats stop and release our tethers, freeing the ropes to dangle and travel just above the water's surface.

A siren sounds from the island, and people flood onto the beaches. They run to boats and to ultralight aircraft.

The boats pull the ultralights aloft, then disconnect them and charge themselves out through the waves. The boats' riflemen take shots at us, while their crews grab for our tethers. It's not hard to destroy these boats with our machine guns.

The ultralight aircraft have a long climb to make. Some of them rise on the updrafts of the island, while others sail out over the water to get behind us. This second group of pilots is going to learn quickly that we are sitting ducks, because our parasails offer no maneuverability. I try to make every shot count, slicing off wings and shooting apart propellers, aided by the tracers in my ammunition.

As we near the island, nests of anti-aircraft guns open fire. I send down a hail of bullets into the trees, then turn my attention to the exposed batteries along the crater's rim. One of my neighboring flyers is shot from the sky.

A few parasailors have reached the target and are dropping their bombs. The resulting explosions send up lashes of flame that engulf some of the lower-flying mercenaries. Those without bombs are incinerated, and those still *with* bombs are incinerated first, then blown to pieces.

Fortunately for me, I'm well out of reach of the flames. I fly high over the crater and opt not to drop my payload until I've reached the far rim. I use my bomb to destroy a gun nest directly beneath me, thereby protecting my slow and undefended escape out to sea.

My tether drags behind me, passing over the crater and through the woods without being captured or snagged. It crosses the beach and out onto the water, where I'm eventually retrieved and hauled down to an awaiting speed boat and fastened to its diving platform.

The James Bond music blares once again, and I'm sped back to the garages and into the blue tunnels.

I'm racing now behind other mercenaries, though they've become difficult for me to see. One of the forward harnesses is empty and smouldering, generating a thick cloud of black smoke in the tube. Soot begins to accumulate on my goggles.

I can just barely perceive that the streaming harnesses have recombined into pairs and are entering Space Mountain's unloading area. My alternate self must have perished, otherwise I'd be heading out for a fourth adventure.

A cast member helps me remove my headgear and lift my shoulder restraints, then he swipes a magnetic card through a slot in my harness and hands it to me.

In the exit hallway, I stop at a video monitor and feed the card into it, where I'm treated to a replay of my three, instead of two, adventures. I'm offered the chance to purchase them.

A final announcement arrives over the speaker system.

> Before your departure today, please be sure to stop at Client Services if, for any reason, your final directive has changed.
>
> While most people want to be reconstituted to their most recent backup, there are certain circumstances that justify other arrangements.
>
> For instance, did someone break your heart recently? Why not choose a backup dated before you met them. Or what about a poor high school experience? Why not go back to the age of thirteen. (Proof of new legal guardianship required.) Or perhaps your mental capacities have grown stodgy and slow, stuck in the worn-out ideas of the past. Why not start all over again at the age of one or two? (Again, legal proof of guardianship required.)
>
> All reconstitution procedures include a full disclosure of the former life being replaced (to be reported to new minors at the age of majority).
>
> Please consider this important decision and make your wishes known.
>
> Thank you for choosing Space Mountain Backups . . . and enjoy your everlasting life.

I emerge from the building to find a small sign posted near

the door. It displays a riddle: "If you wake up as someone else tomorrow, how will you know it?" The answer is hidden under a flap. "You won't, because every neuron in your brain will tell you who you really are and who you've always been."

"Scary, Bert. If I woke up as you, I'd be too dumb to want myself back." −Dad

"Absolutely correct, Dad. The one and only soul of the universe is everywhere and in everything—we all share it. You and I are just two of its many patterns that don't want to be replicas of each other." −Bert

"Congratulations, Bert. The meaning of the word 'soul' is changing once again, this time transcending the individual and extending to all of creation. (I've duly noted the new usage in the margins of my Oxford Dictionary.) I wonder what effect this will have on the Social Contract."

−Uncle Albert

CHAPTER 10

Circle-Vision 360
(formerly *Monsters, Inc. Laugh Floor*)

TYPE: surround-screen film presentation
DURATION: 20 minutes
CAPACITY: 1,000+
HEIGHT REQUIREMENT: none
LOADING SPEED: fast
AVERAGE WAIT: 20 minutes or less
RECOMMENDED AGES: 13 and up
DESCRIPTION: Guests stand alongside handrails within a pavilion, where nine screens provide panoramic viewing of a film called *Spirit Mountain*.

The theater lights dim until the entire room is dark. Then, an unseen orchestra begins its warm-up, producing sounds that ebb and flow. Three taps of the conductor's baton brings silence, and an oboe plays a Concert A.

One of the screens displays a white line, which wiggles to the oboe's note. As more instruments join in, other lines appear, and soon, all of the screens are alive with chaotic scribbles.

Another tapping by the conductor brings pause, then the music begins. The lines on the screens perform a dance of elegant patterns, overlaid by the voice of a narrator:

> The stuff of the universe—call it dark matter or dark energy—plays like the grandest of symphonies . . .

where sounds form into notes, notes form into harmonies, harmonies grow into melodies, melodies
carry a theme, and themes whip up emotions . . . and
all from simple sounds. The stuff of the universe is
no different: the ether forms into atoms, the atoms
form into molecules, the molecules grow into instructions, the instructions carry life, and life whips up
emotions . . . and all from a simple ether.

This layering is the basis of all creation. No layer
can emerge without all of its subordinate layers
remaining intact. Note that emotions require life;
life requires instructions; instructions require molecules; molecules require atoms; and atoms require
the ether. Each higher level depends on the viability
of all those beneath it, but not vice versa. Creation
is directional. Humankind is on a natural path.

The story for today follows Emotion on its
extended journey. And much of the trail you're about
to see was forged long ago by your ancestors. You'll
likely recognize a good deal of it yourself, given that
you've already trekked much of it during your own
lifetime. But keep in mind that some of the more
advanced trails may be unknown to you.

The journey begins on a desert salt flat.

The music continues, but now the screens have become
bright white. I can just barely make out a horizon line.

The worldview of early humans was washed out like
this. These creatures had no ability to look inward
and observe their own thoughts. They wore their
emotions on an invisible shirtsleeve, reacting instinc-

tively to every stimulus. They were unaware of their existence on these *Archaic Salt Flats,* yet they lived here just the same.

The orchestra fades, leaving behind a primitive flute. The images of the ground turn gradually from a white desert to a lush green everglade, where a thick fog blots out the sky and reduces visibility to a few meters. The terrain now passes more slowly.

Here in the *Magic Swamp,* memories of the past and anticipations of the future are collapsed into the *here and now.* They arrive as sensations and drive natural impulses, though they are absent of any conscious attendant.

The great civilizations of ancient Egypt and the early Americas dwelled in this *Magic Swamp.* They achieved stunning levels of social complexity, but they did it without any clear self-knowledge.

The gentle sounds of a stringed instrument replace those of the flute. The cameras climb from the swamp and enter a foggy entanglement of trees, vines, and underbrush.

Welcome to the *Jungle of the Power Gods.*

Electrification overcomes the acoustic strings, and the music launches into the opening sequence of Guns N' Roses' *Welcome to the Jungle.*

To hell with the harps and lyres of the time. Let's unleash the terror and the ecstasy that only modern strings can convey. The *Jungle of the Power Gods* is a dream and a nightmare all in one. The mist has thinned; the path forward and backward can now be seen. History plots on a timeline, and time is real and on the march. It can't be stopped . . . meaning death is inevitable.

Yes, death!

Death is the concept that opposes *life;* you can't know one without knowing the other. The gods of the ego kill with a vengeance. . . .

The music blends into a later part of the song, where high-pitched pinging sounds cry against the thumping of an electric bass. Guitar is added, and the lead singer screams, "You know where you are? You're in the jungle, baby. You're gonna die!"

The music swells, then suddenly drops off to the lone stringed instrument of before.

But a little self-control is in order here.

A fairy-tale picture appears alongside the cut trail, and propped above it is a carved wooden arrow that points forward. The arrow bears the inscription, *"Works and Days"* —*Hesiod.*

Isn't that image inspiring? —A jungle free of fog, vines, and undergrowth, where deciduous trees tower like pillars over a carpet of last year's leaves. That's the dreamworld from the *Jungle of the Power Gods.*

But such ideal images are seldom true in every detail. The coming *Mythic Forest* will be littered with fallen trees and thorny brambles, and the ground will be dangerously uneven and sloped upward.

The scene changes to what the narrator described. The *Mythic Forest* is a dense and treacherous woods, with a walking path that is marked by religious statuary.

> The *concrete operational* mind is governed by perceptions and experiences. It can't step away from an idea and evaluate it objectively. The holy scriptures are a literal account of history — true word-for-word — and the separation of church and state is meaningless.

Another easel along the path shows a picture of the forest giving way to scattered pine trees, where a perfect stairway of stone climbs easily up the slope. The carved wooden arrow reads, *"Republic"* — *Plato.*

> Once again, ideal images tend to fall short of reality. The *Rational Pines* occupy a hazardously steep incline, with jagged rock outcroppings instead of smooth stairs. Note that the trail is clearly marked by mechanical gadgetry, like clocks and printing presses and . . . oh yes, one more religious statue.
>
> Mother Teresa made it to the *Rational Pines,* and from here, she grappled with her own doubts about God for four decades until her death.
>
> The *Rational Pines* offer a huge leap in perception. *Logic* and *reason* become the toolkits for invention, and the existence of God falls into question. The traveler through this region is a self-made individual.
>
> Note, however, that this traveler remains hemmed in by tree cover. The view forward and backward continues to be quite limited.

A third easel presents a picture of a sheer rock face, basking in the sunshine. The wooden arrow above it is inscribed, *"Utopia" — Thomas More.*

> The first people to reach the *Pluralist Cliffs* found the rocks to be without hand-holds or foot-holds. Every inch of their climb was hard won, and each climber had to affix the necessary anchors for his or her own personal ascent. Some of these anchors, as you can see, still bear the tie-dyed ribbons they were decorated with.
>
> If you listen carefully among these cliffs, you can hear the lingering echoes: "Free-to-be-you-and-me. . . . Let's celebrate diversity. . . . I want to hear every voice. . . . Spare me your hierarchies."
>
> These people of the *Pluralist Cliffs* seldom look back at the landscape below. But on those rare occasions when they do, they fail to recognize that their superior vantage point owes to the arduous path they've already traversed. They deny the very hierarchy they sit at the top of.
>
> Yet, there's one more easel to be found. This one is strapped to a protruding rock. Its picture shows a brilliant white snowcap above the cliffs, but its attached arrow is made unreadable by the sun's glare.

The cameras advance up the mountainside, and the front three screens of the theater turn white with snow, reminiscent of the earlier salt flats. The other six screens show the raw beauty of the countryside below.

> Few climbers have attained the lofty reaches of the *Integral Snowcap.* Those who have, however, can

finally make sense of the lower elevations: the *Archaic Salt Flats,* the *Magic Swamp,* the *Jungle of the Power Gods,* the *Mythic Forest,* the *Rational Pines,* and the *Pluralist Cliffs.*

These climbers understand that children must start from the beginning and move through every lower terrain. They also understand that the world's current events are dominated by the three regions immediately below them.

Nations of the *Mythic Forest* are ruled by despot kings, who crush opponents, purge dissenters, and seek a glorious empire. Their citizenry is generally capable of committing genocide.

The corporate states of the *Rational Pines* are far more civilized. They tend to avoid conflicts unless oil or shipping or some other national interest is threatened. Then, they deliver an ultimatum to the offending party, rushing in with superior weaponry and installing a more favorable government if the leadership refuses to yield.

The pacifists of the *Pluralist Cliffs* cry for restraint. They want all parties to sit and negotiate a peaceful settlement. And when that doesn't work and war breaks out, they insist that the replacement governmental system be a republic.

But the climbers of the *Integral Snowcap* see the folly of the *Rationals* and the *Pluralists* in handling the *Mythics.* The violence of the *Rationals* simply breeds more violence down the road, and the *Pluralists'* call for peace is as naïve as their imposition of a democracy on a society that's not ready for one.

And what about religion?

The clergy of the *Mythic Forest* don't tolerate an

atheist or an agnostic of the *Rational Pines* within their ranks. And even the clergy of the *Pluralist Cliffs* sometimes find it difficult to make peace with their earlier traditions. Consider, for example, the Dalai Lama and his conflicted statements about sexuality. But travelers of the *Integral Snowcap* know that the true spiritual path includes many terrains, and that each one is a necessary crossing for growth to continue.

The snowcap is the place where the long-standing problems of the lower levels can finally be solved. And the solutions—you won't believe—are entirely elegant and simple.

The camera climbs toward the mountain's summit, and the forward horizon comes into view.

But the intellectual trickle-down is progressing too slowly. The civilizations of the *Mythic Forest* are acquiring the advanced technologies of the *Rational Pines* and the *Pluralist Cliffs*. If the world's intellect doesn't rise in a hurry, the coming struggle will likely be all or nothing. The future seems geared toward two possible outcomes.

The music changes to Louis Armstrong's *What a Wonderful World,* and the current pictures dissolve into a montage of lovely images. Text floats intermittently across inspiring scenes: "The preferred path is a world of peace and prosperity for all people, where the oceans come alive again . . . where skies are free of pollution . . . where the global climate is safely in balance . . . where clean energy is cheap and abundant . . . where cities and towns radiate with charm and vitality . . . where deserts bloom with life . . . where people transcend their own biologies . . . and where humankind unites on a quest for survival and ultimate meaning. . . . What a wonderful world." The song ends, and all nine screens fade to black.

But the more likely scenario is not so rosy.

An acoustic guitar plucks out some individual notes, beginning the song *Handlebars* by the Flobots.

Each screen shows a human face of a different nationality. The faces are all male, and they rotate away from the cameras. The nine men stand on a hillside overlooking the skyline of a city.

They ride their bicycles into the downtown and through the tree-lined streets of a college campus, stopping at a plaza to synchronize the countdown timers on their cell phones. Each of them hugs the others, then they go their separate ways.

The screens cut to video footage of civil rights demonstrations—both past and present—where police and militia have turned violent against peaceful protesters. Water and teargas escalate to billy clubs, rifles, and tanks, littering the ground with the dead and wounded in pools of blood. Some of the films are black and white. Others are in color.

The nine men travel on airplanes and trains and buses, sharing videos and text messages. One video clip shows a scene from the movie *Monty Python and the Holy Grail,* and the accompanying text message reads, "Bring out your dead." Another clip shows the Road Warrior firing an automatic weapon into a lineup of naked men and women, who then crumple into a trench. Charlton Heston wears a loin cloth and kneels on a beach in front of a half-buried Statue of Liberty. "We finally really did it," reads the text message. "God damn you all to hell!" The students smile.

Images of violence against women are shown opposite the fashion magazines that glamorize their figures. Nuns and other fully enshrouded women are shown opposite the world's neglected and battered children. Closed-captioning across the screens delivers quotations that blame women and young girls for their own rapes, and which further deny that "spousal rape" is a crime. Benny Hill smirks, then chases a gathering of scantily-clad women around the screens.

World leaders of the twentieth century gesticulate behind microphones, as their populations listen and their armies go to war. Bodies are bulldozed, and Major Kong rides the bomb in *Dr. Strangelove*.

New leaders replace the old, and the twenty-first century begins. Inspectors walk with clipboards through a facility, and bombs fall on Bagdad. A rocket is test-fired.

Each of the nine men finds a duffel bag left for him on a doorstep or in a public locker or in the backseat of a car. One of the men opens his bag and finds it loaded with cash and a plane ticket. The remaining eight travel by various means into the downtowns of large cities. They emerge from stairways onto the rooftops of skyscrapers and joyously survey the spectacular scenery around them.

The screens become filled with the orphans of Africa, sick and impoverished, while the wealthy of the world take inoculations in sterile efficiency.

Every execution and assassination that has ever been captured on film seems to be showing at this moment.

Video footage transmitted from *smart bombs* replays with the final moments before impact.

Black and white images show the charred and disfigured remains of people yet alive in a destroyed city.

The man with the plane ticket touches down on a tropical island and finds his way to a seaside bar. There, he sits where he can watch a television screen and also the women on the beach. His cell phone displays the countdown: ten, nine, eight, . . .

The men on the buildings rise from their meditations, but the calm on their faces can't hide the menace behind their eyes. They move to the ledges and begin swinging their duffel bags.

With screams of hell-bent fury, they release the bags into the air and raise their fists. One man does a swan dive off his building.

Every screen but one erupts into fire.

The music falls away, back to the simple notes of the song's introduction, and each screen displays a different one of the silent nuclear explosions from the end of *Dr. Strangelove*.

Text creates a ring around the now darkened room. "That portion of the soul we call Earth is now climbing Spirit Mountain. Will it rise on through the ages? . . . Or will it fall back to start anew? . . . Which outcome will you cause?"

The house lights come on, and the theater doors swing open. The soundtrack restarts with the Guns N' Roses tune of earlier.

"Everyone probably thinks they're near the top of Spirit Mountain. Who could possibly convince them that they're wrong?"   –Uncle Albert

CHAPTER 11

The Tomorrowland Oracle
(new show)

TIME: before the *Wishes* fireworks show
DURATION: 15 minutes
RECOMMENDED AGES: 7 and up
DESCRIPTION: This nightly theatrical performance takes place
   on the rooftop of Space Mountain and features acrobats,
   laser lights, pyrotechnics, and more.

I remain in Tomorrowland for the rest of the day, catching
a taping of *Sculpt Me Beautiful* on the Tomorrowland Stage
and enjoying a second meal at Cosmic Ray's Café. Tonight's
dinner entertainment was *The Heavenly Bodies*—an acting
duo that dresses as the planets Mars and Venus to perform
humorous skits, showcasing the hormonal differences between
males and females.

It's after dark when I leave the restaurant and skirt the
many guests now gathering on the plaza. I join some others,
who have found the secluded footpath connecting Tomorrowland with Toontown. We are all awaiting the start of the
nighttime spectacle on Space Mountain.

The Mountain's familiar dome features raised seams that
ascend to a crown of stylized spires. This outer shell is currently aglow in soft lighting.

A remote pulsing sound emerges and gains in volume, coin-
ciding with an ever-brighter flashing of the lights surrounding

the dome. The crowd gives rise to a great cheer.

The lights chase back and forth across the rooftop, then a minor explosion on the crown sends a billow of smoke into the air between the spires.

When the smoke finally clears, a human figure is standing atop the building, wearing a dark robe that bristles with strands of electricity. I can hear the voice of an old woman humming quietly.

The lower span of Space Mountain's rooftop displays the start-up screen for the video game *Guitar Hero,* and an amplifier crackles as someone works swiftly through the menus. *Expert* is the chosen difficulty level. *Eruption,* by Van Halen, is the chosen song. A fret board tips into view, scrolling like a conveyor belt down the dome and carrying a cargo of colorful dots. There's a flurry of drumbeats, then an electric guitar slams into an initial chord, as the dots drop off into rings of fire. A spotlight finds the robed figure, and she appears to be playing a small guitar.

For ninety ear-ripping seconds, Tomorrowland sizzles under a fire-storm of notes, eventually reaching its conclusion in a trailing echo of distortion. Audience approval swells to fill the void left behind.

The cloaked figure raises the guitar in one hand. "You kids got nothin' on me," says the thin voice of an elderly woman. "Turn down that music," she mocks a parental tone.

She sets the guitar at her feet, then sweeps her arms through the sky like the oracle of this morning. Her right arm slowly fixes its aim down into Tomorrowland. "You!" she says, "—each and every one. . . . You are the makers of tomorrow. . . . You are the curators of joy and sorrow. . . .

Don't look to the gods to spare you your plight,
You are the ones who must own the whole fight.

Technology is racing ahead of your brain.
Unless you get wise, things will get more insane.

So listen up now, and listen up good.
I'm sharing one thing of the things that I should.

Your problems are many . . . so many in fact,
That too many answers means failure to act.

So here's just one task. . . . I'm keeping it small.
Do this one thing and you'll help solve them all.

Corruption in government is the seed of your trouble.
Stop this corruption, and proceed on the double.

You'll discover it's tragic, though piggishly funny.
Just look for the pork and follow the money.

It's *campaign finance* at the top of your list.
If you're feeling some rage, please show me your fist.

Your captains of business will fight, tooth and nail.
Their government puppets will work to prevail.

The sound of a helicopter rises from the woods behind
Tomorrowland. It hovers beside Space Mountain with a spotlight trained on the old woman. But rather than turning to
face it, she holds her gaze down into Tomorrowland.

"Surrender," shouts a male voice through a speaker from
the helicopter. "We don't want to shoot."

A line of men, dressed all in black, run along the lower
eave of Space Mountain. Each of them finds a space between
the seams of the roof and begins to climb.

The old woman reaches for her guitar and punches up another song. She uses the guitar's neck to steer the projected fret board around the roof's surface.

The acrobatic climbers jump and swing themselves out of the way of the descending notes, but whenever they are hit, they are knocked back down to the eave to restart their climb.

As these ninja-types encroach on the crown of the building, the woman returns to her verse:

> Your captains of industry will fight, tooth and nail.
> It is *you,* with your ballots, who must strive to prevail.
>
> So bring me your laws when you think they're all done.
> I'll grade them and score them and tell you who's won.
>
> That's right; it's a contest. I work in four nations.
> I welcome all states to submit their creations.
>
> Who will aspire to be head of the class?
> A move to *Step Two* is the prize if you pass.

The men in black are nearing the summit now. A few of them travel along the ledge of the lower ring.

> I can't do it alone, though, and neither can you.
> We must all work together to see this thing through.
>
> To win by the vote is the civilized way.
> So please raise your hands and shout 'Ballots, hurray!'

Two of the men have snuck up behind her and are now holding a sack open above her head.

"Ballots, hurray!" comes the reply from the audience.

Blue laser lights streak from every rooftop in Tomorrowland and converge on the old woman's guitar. The lights stream through it and spill out onto the building through the notes, zapping the men in black and causing them to cascade like dead flies down the surface of the dome.

"You leave us no choice," announces the man from the helicopter. A missile fires from beneath the cockpit.

The missile swings wide, deflected by the web of blue lasers. It encircles Space Mountain and returns to the helicopter, clipping the main rotor on its way past. The craft falls backward into a tailspin, then sinks below the trees. There's a crash, then an explosion, and a fireball rolls into the sky. I can feel the heat from where I stand.

The old woman holds her guitar overhead and laughs. "Take *that*," she shouts. "COMMAND THE VOTE!"

Then, she's gone in a puff of smoke.

The show is over, and the lights of Space Mountain return to normal.

It's time for me to be going home now, so I head for the main exit of Tomorrowland. There, on the back of an elaborate entrance sign, is a special farewell message: "Stop living for today and start living for tomorrow—turn conventional thinking on its head. Tomorrow just got a whole lot brighter."

"Politicians play by the rules we set for them. When the rules are ridiculous, so are the outcomes."   −Bert

"I loved the whole thing, Bert . . . and so did your father."
                                                    −Mom

Epilogue

It doesn't bother me that the experts in my family dismiss me as naïve for my ideas. I know that the truth is more powerful than a lie — good nutrition makes a body healthier, and wise investing makes a person wealthier — so the issue of my humiliation or vindication will surely resolve itself in due time, regardless of my uncredentialed status.

Thus, with zero reputation to lose for my statements, I might as well say it all.

The entire sweep of human experience — past, present, and future — now makes perfect sense to me. The human species is following a natural path of progress that is accelerating it toward an inflection point, set to occur this century. If humankind can somehow survive the singularity and its surrounding shakeout, then a Golden Age of Consciousness awaits.

One thing seems certain, though. The Benjamin Franklins of the world must soon take their Pinocchios by the hand and lead them to a better tomorrow. The Lantern of Truth must light the way.

Further Reading
(updated)

Natural Capitalism: Creating the Next Industrial Revolution by Paul Hawken, Amory Lovins, & L. Hunter Lovins (1999)

The First Idea: How Symbols, Language, and Intelligence Evolved From Our Primate Ancestors to Modern Humans by Stanley I. Greenspan & Stuart G. Shanker (2004)

The Art and Science of Low Carbohydrate Living: An Expert Guide to Making the Life-Saving Benefits of Carbohydrate Restriction Sustainable and Enjoyable by Jeff S. Volek & Stephen D. Phinney (2011)

The Primal Blueprint by Mark Sisson (2009)

The Complete Guide to Fasting: Heal Your Body Through Intermittent, Alternate-Day, and Extended Fasting by Dr. Jason Fung (2016)

Body by Science: A Research-Based Program for Strength Training, Body Building, and Complete Fitness in 12 Minutes a Week by Doug McGuff & John R. Little (2009)

Mastering the Market Cycle: Getting the Odds on Your Side by Howard Marks (2018)

Jimmy Stewart is Dead: Ending the World's Ongoing Financial Plague with Limited Purpose Banking by Laurence J. Kotlikoff (2010)

The Singularity Is Near: When Humans Transcend Biology by Ray Kurzweil (2005)

A Brief History of Everything by Ken Wilber (1996)

The Origin of Consciousness in the Breakdown of the Bicameral Mind by Julian Jaynes (1976)

Republic, Lost: How Money Corrupts Congress – And a Plan to Stop It by Lawrence Lessig (2011)

Please support passage of the *American Anti-Corruption Act*. You can learn more about it and how to unrig our political system by watching Jennifer Lawrence at https://represent. us/unbreaking-america/.

Other titles by L.N. Smith

Grand Unification and The New Look of the Atom (2004)

Sunrise Over Disney (2011)

www.ingramcontent.com/pod-product-compliance
Lightning Source LLC
Chambersburg PA
CBHW022124280326
41933CB00007B/537